STEPHEN EMMOTT

TEN BILL1ON

Stephen Emmott is head of Computational Science at Microsoft Research. He leads a broad scientific research program, at the center of which is an interdisciplinary team of new kinds of scientists, and a new kind of laboratory, in Cambridge, England, pioneering new approaches to tackle fundamental problems in science. His lab's research spans from molecular biology, immunology, and neuroscience, to plant biology, climatology, biogeochemistry, terrestrial and marine ecology, and conversation biology, as well as the new fields of programming life and artificial photosynthesis. Stephen is also Visiting Professor of Computational Science, University of Oxford; Visiting Professor of Biological Computation, University College London; and Distinguished Fellow of the UK National Endowment of Science, Technology and the Arts.

TEN
BILL10N

TEN BILL1ON

STEPHEN EMMOTT

VINTAGE BOOKS
A Division of Random House, Inc.
New York

A VINTAGE BOOKS ORIGINAL, AUGUST 2013

The Cataloging-in-Publication Data is on
file at the Library of Congress.

Vintage ISBN: 978-0-345-80647-5

Book design by Heather Kelly

www.vintagebooks.com

Printed in the United States of America
10 9 8 7 6 5 4 3 2 1

TEN

BILL10N

This is a book about us.

It's a book about you, your children, your parents, your friends. It's about every one of us. It's about our failure: failure as individuals, the failure of business, and the failure of our politicians.

It's about the unprecedented planetary emergency we've created.

It's about the future of us.

Earth is home to millions of species.

Just one dominates it. Us.

4F
永乐生活电

Apartment buildings in Shanghai

Our cleverness, our inventiveness, and our activities have modified almost every part of our planet. In fact, we are having a profound impact on it.

Indeed, our cleverness, our inventiveness, and our activities are now the drivers of every global problem we face.

And every one of these problems is accelerating as we continue to grow toward a population of ten billion.

In fact, I believe we can rightly call the situation we're in right now an emergency—an unprecedented planetary emergency.

This is the reason I have written this book.

I am a scientist.

I lead a lab, in Cambridge, England, which is home to a unique collection of amazing young scientists. We conduct research into complex systems, including the climate system and ecosystems, as well as the impact of us humans on the earth.

Science is ultimately about *understanding*. And this is what we try to do: to understand the earth's climate, and the behavior of the earth's terrestrial and marine ecosystems—from its microbial communities to its forests—and to predict how these vital planetary systems will respond to change.

Change caused by us.

We humans emerged as a species about 200,000 years ago. In geological time, that is really incredibly recent.

Just over 10,000 years ago, there were one million of us.

By 1800, just over two hundred years ago, there were one billion of us.

By 1960, fifty years ago, there were three billion of us.

There are now over seven billion of us.

By 2050, your children, or your children's children, will be living on a planet with at least nine billion other people.

Sometime toward the end of this century, there will be at least ten billion of us. Possibly more.

How did we get to where we are now?

World Population (Billions)

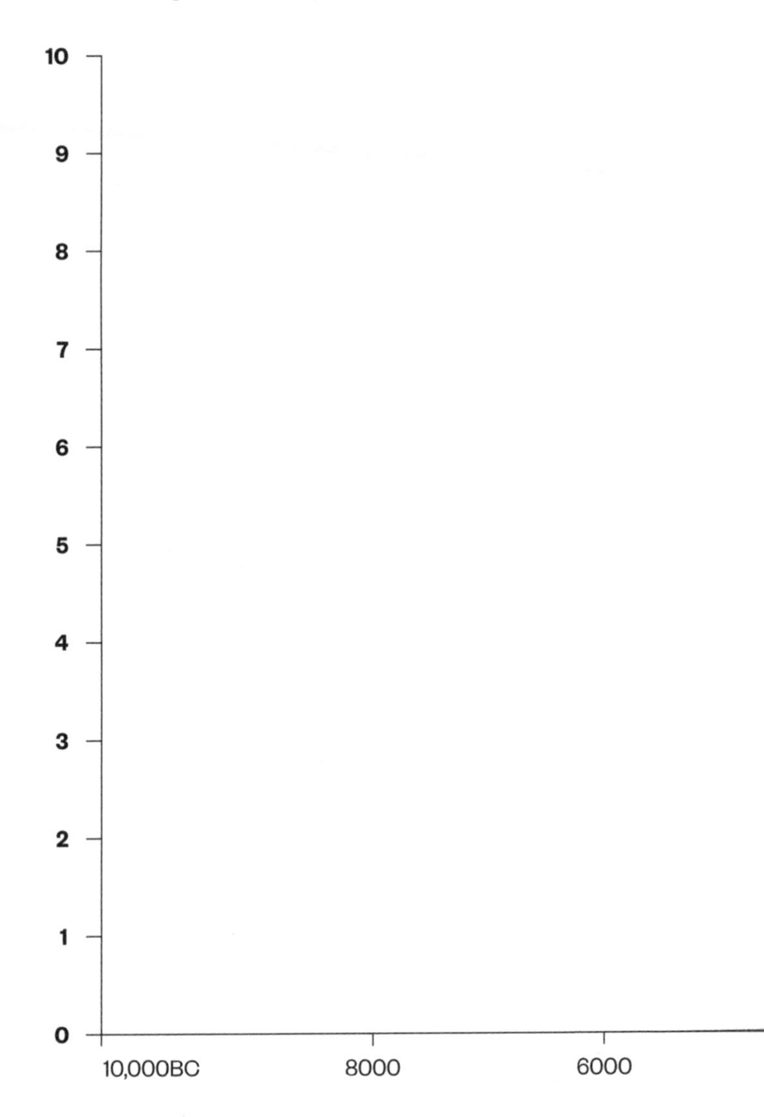

Human Population from 10,000 B.C., projected to 2100. Data from 1950 from United Nations Department of Economic and Social Affairs, Population Division: World Population Prospects, 2011.

4000
2000
1 AD
1000
2100

We got to where we are now through a number of civilization- and society-shaping "events"; most notably, the agricultural revolution, the scientific revolution, and—in the West—the public-health revolution.

These events have fundamentally shaped how we live, and have fundamentally shaped our planet. Their legacy will continue to shape our future. So we need to look at our growth and activities through the lens of these developments.

By 1800 the global population had reached one billion.

One of the principal reasons for this growth was the invention of agriculture. The "agricultural revolution" enabled us to go from being hunter-gatherers to highly organized producers of food, and allowed our population to grow.

A useful way to think of the development and importance of agriculture is in terms of at least three agricultural "revolutions." The first took place over 10,000 years ago. This was the domestication of animals and the cultivation of plant types.

The second agricultural revolution was between the fifteenth and nineteenth centuries. This was a revolution in agricultural productivity and the mechanization of food production.

The third happened between the 1950s and 2000s; the so-called "green revolution."

But there's another story here: the start of a fundamental transformation—of land use—by humans.

One hundred and thirty years later, we had grown to two billion.

It was 1930. The impact of another revolution—the industrial revolution—was being felt. The world was being transformed by manufacturing, technological innovation, new industrial processes, and transportation.

The continuing expansion of agriculture and the revolution in public health enabled us to continue to grow—rapidly.

But there's another story here too: the start of our lethal addiction to coal, oil, and gas as our principal sources of energy.

Thirty years later, we had grown to three billion.

It was 1960, and we were in the middle of a food revolution. There were more of us. Far more of us. We needed more food. Far more food. More than the established agricultural system could provide.

What became known as the green revolution provided this extra food.

It did so through:

> The industrial-scale use of chemical pesticides, chemical herbicides, and chemical fertilizers;
>
> an unprecedented expansion of land use;
>
> and the wholesale industrialization of the entire food production system. This included the industrialization of raising and harvesting animals for food, from the rise of industrial-scale "factory fishing" fleets to battery farming of pigs, poultry, and beef.

This revolution came at a huge cost to the environment, in terms of:

loss of habitat;

pollution;

overfishing.

It also set in motion an unprecedented decline of species and the start of the degradation of entire ecosystems.

By 1980, twenty years later, there were four billion of us on the planet.

The green revolution had produced much more food. That made food cheaper.

In turn, that meant we had more money to spend. And we had started to spend it on "stuff": televisions, video recorders, Walkmans, hair dryers, cars, and clothes. And we also started to spend it on vacations. Far more vacations.

At the center of this spending spree was the astonishing growth of transportation.

In 1960 there were 100 million cars on the world's roads—by 1980 there were 300 million.

Freeways, Los Angeles

With this came a massive expansion of road networks—carving up entire countries, further increasing loss of habitat for other species.

In 1960 we flew 62 billion passenger miles. In 1980 we flew 620 billion passenger miles.

Global shipping grew at a similarly astonishing rate. All of the stuff we were buying, plus all of the food we were consuming, plus all the raw materials and resources required to make everything was being shipped around the world.

Just ten years later, in 1990, there were five billion of us.

By this point, initial signs of the consequences of our growth were starting to show.

Not the least of these was on water.

Our demand for water—not just the water we drank, but the water we needed for food production and to make all the stuff we were consuming—was going through the roof.

But something was starting to happen to water. Back in 1984, journalists reported from Ethiopia about a famine of biblical proportions caused by widespread drought.

That, it seemed, was “over there,” in Africa. Except that it wasn’t just happening “over there,” in Africa. Unusual drought, and unusual flooding, was increasing everywhere: Australia, Asia, Europe, the United States.

Water, a vital resource we had thought of as abundant, was now suddenly something that had the potential to be scarce.

By the year 2000, there were six billion of us.

By this point it was becoming clear to the world's scientific community that the accumulation of CO_2, methane, and other greenhouse gases in the atmosphere—as a result of agriculture, land use, and the production, processing, and transportation of everything we were consuming—was changing the climate. And that, as a result, we had a serious problem on our hands.

Nineteen ninety-eight had been the warmest year on record.

The ten warmest years on record have occurred since 1998.

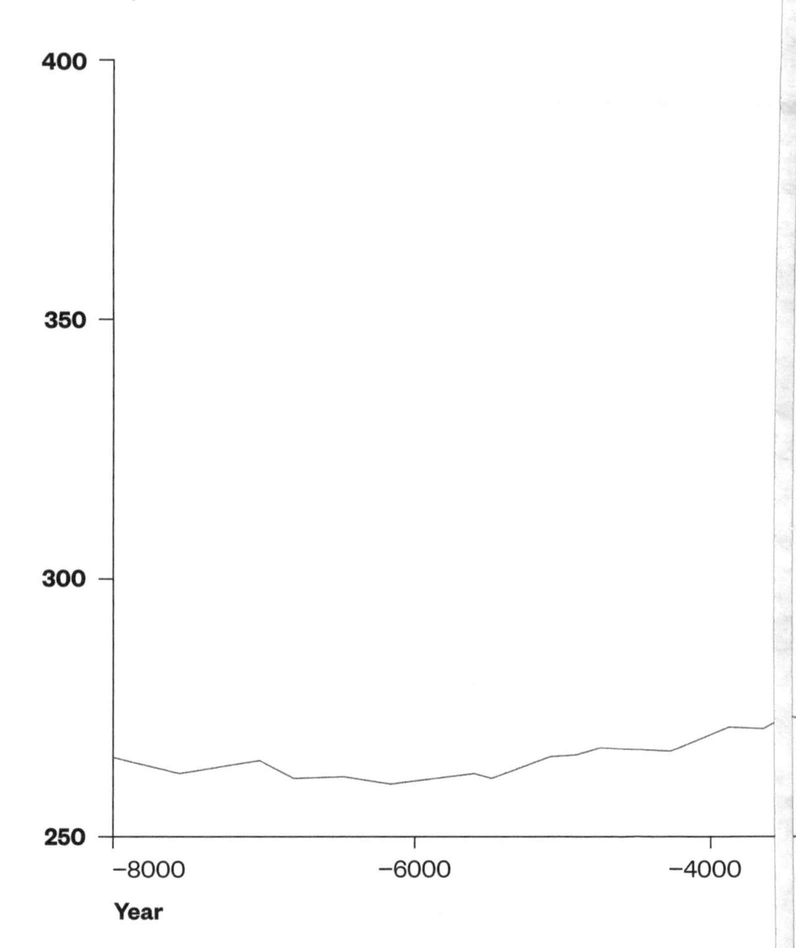

CO_2 concentration is increasing. CO_2 concentrations have increased from 280 parts per million (ppm) at the start of the industrial revolution to 400 ppm in 2013 (measured for the first time on May 4, 2013). To keep a global average temperature rise to 2 degrees C, we would need to limit CO_2 concentrations to 425–450 ppm. We will not meet this target. A more realistic target, if we were committed to actually tackling climate change—which we're not—would be 550 ppm. Even this assumes that the planet's vegetation and oceans continue to

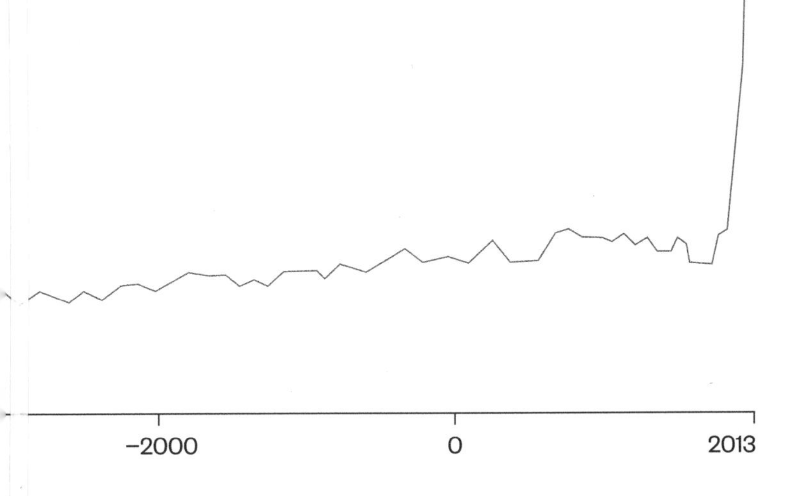

do us the massive favor of acting as enormous carbon "sinks." And then we are into the territory of a 4-degree rise in average temperature. Adapted from the Scripps Institute CO_2 program, "Climate Change 2007: The Physical Science Basis," Working Group 1 contribution to IPCC, *Fourth Assessment Report,* CUP 2007; C.M. MacFarling Meure, et al. "Law Dome CO_2, CH_4, and N_2O ice core records extended to 2000 years BP," *Geophysical Research Letters* 33, 14 (2006)

We hear the term “climate” every day, so it is worth thinking about what we actually mean by it.

Obviously, “climate” is not the same as weather.

The climate is one of Earth’s fundamental life support systems, one that determines whether or not we humans, and millions of other species, are able to live on this planet. It is generated by four components:

> the atmosphere (the air we breathe);
>
> the hydrosphere (the planet’s water);
>
> the cryosphere (the ice sheet and glaciers);
>
> the biosphere (the planet’s plants and animals).

By now, our activities had started to modify every one of these components. Our emissions of CO_2 had started to modify our atmosphere. Our increasing water use had started to modify our hydrosphere.

Rising atmospheric and sea-surface temperatures had started to modify the cryosphere, most notably in the unexpected shrinking of the Arctic and Greenland ice sheets.

Our increasing use of land, for agriculture, cities, roads, mining—as well as all the pollution we were creating—had started to modify our biosphere.

Or, to put it another way: We had started to change our climate.

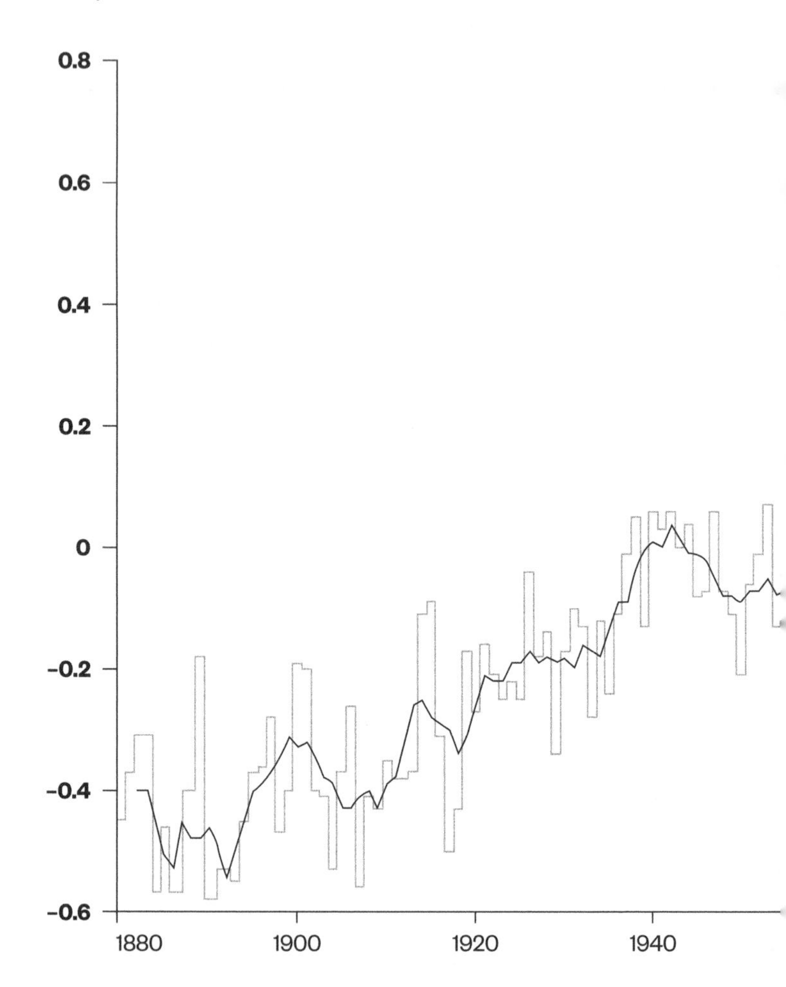

The world is warming. Although the past decade has seen a slowdown in global average atmospheric temperature increase, the long-term trend is a continuing rise in global average temperature: a consequence of increasing levels of atmospheric CO_2. An inescapable fact of physics and chemistry is that

increasing CO_2 in the atmosphere will lead to increasing atmospheric temperatures. The black line is five-year average, the gray line is annual. Adapted from J.E. Hansen, et al., Goddard Institute for Space Studies, 2012: http://cdiac.ornl.gov/trends/temp/hansen/graphics.html

Global Ocean Warming

Heat content, 10^{22} Joules

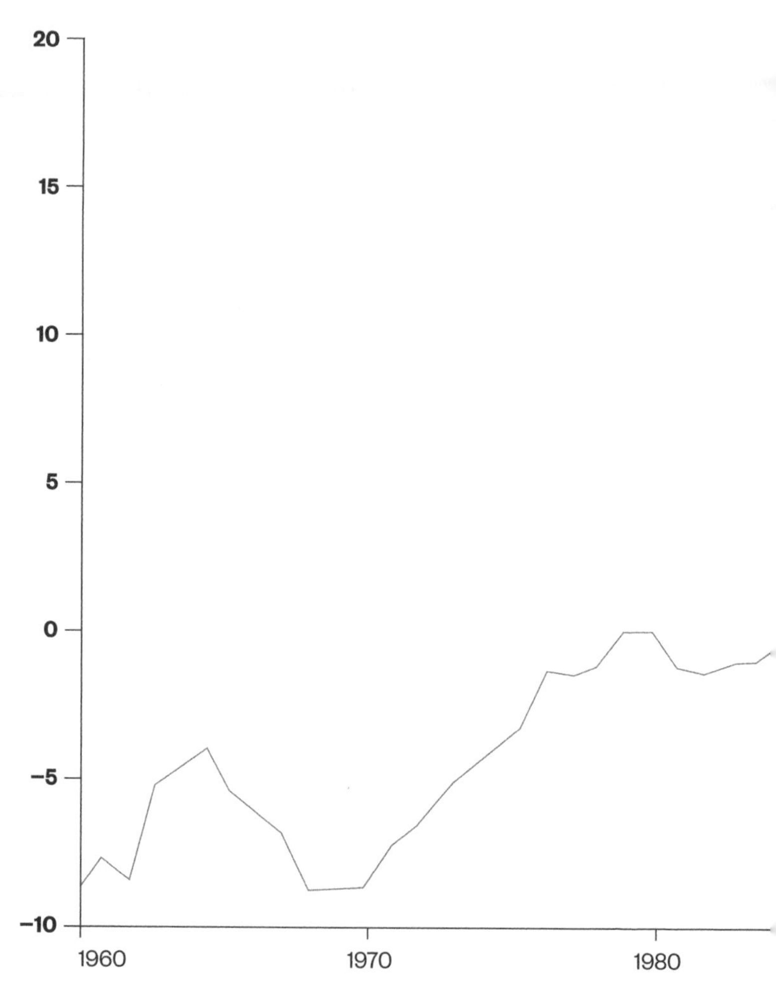

Much of the global warming over the past decade has gone into heating up the planet's oceans, as these data show. When we talk about "global warming" we really need to look at the whole system—the Earth System. At this level, warming continues to accelerate. Data show Global Ocean Heat Content

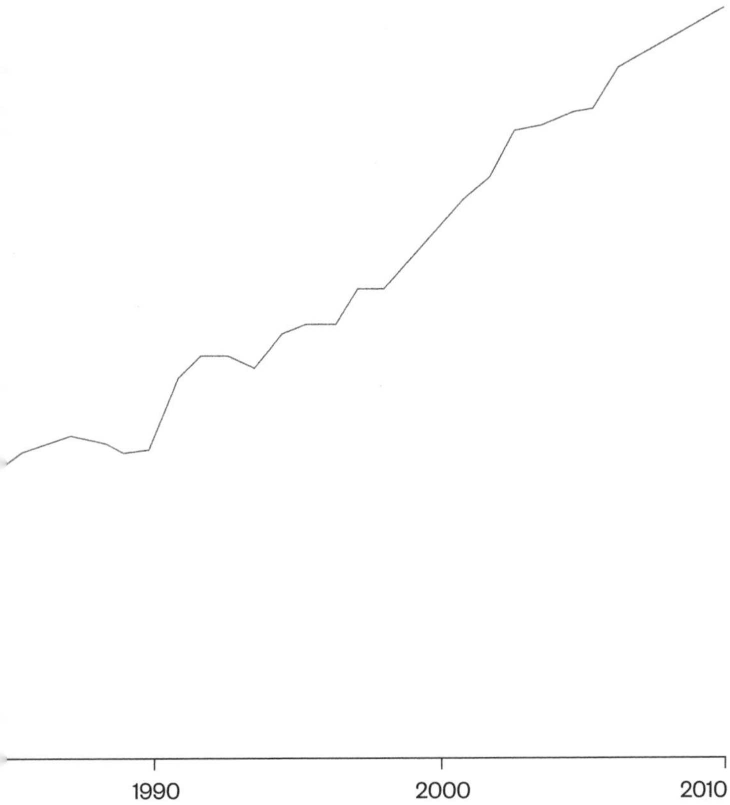

1990

2000

2010

(OHC) for 0–2 km depths. Adapted from data from the NOAA Ocean Climate Laboratory; S. Levitus et al., "World ocean heat content and thermosteric sea level change (0–2000 m), 1955–2010," *Geophysical Research Letters*, 39, 10 (2012)

In just the past twelve years, we've grown by yet another billion.

There are now more than seven billion of us on Earth.

As our numbers continue to grow, we continue to increase our need for far more water, far more food, far more land, far more transportation, and far more energy.

As a result, we are now accelerating the rate at which we're changing our climate.

In fact, our activities are not only completely interconnected with, but are now also interacting with, the complex system we live on: Earth.

It is important to understand how all this is connected.

An increasing population accelerates the demand for more water and more food.

Demand for more food increases the need for more land, which accelerates deforestation.

Increasing demand for food also increases food processing and transportation.

All of these accelerate the demand for more energy.

This then accelerates greenhouse gas emissions, principally CO_2 and methane, which further accelerate climate change.

As climate change accelerates, it increases stress on water, food, and land. And at the same time, an increasing population also accelerates stress on water, food, and land.

In short, as population increases, and as economies grow, stress on the entire system accelerates sharply.

This is where we are right now.

We need to take a closer look at what's happening right now—today—with this highly interconnected system that we rely upon, and which we are rapidly changing. Because doing so is critical to understanding where we are heading.

Right now nearly 40 percent of the entire ice-free land surface of our planet is being used for agriculture.

The remaining land comprises:

1. The Sahara and large parts of Australia—none of which are usable for agriculture. Siberia and other tundra, which are not usable for agriculture—yet.
2. The places where we live: cities and towns—and the spectacular infrastructure that comes with them—road and rail networks, airports, ports.
3. Protected areas, such as national parks.
4. Land being used for extraction of the earth's finite resources: coal, oil, gold, rare earths, iron, copper, zinc, minerals, and phosphates.
5. Managed forests, which are used for timber production.

Texas

The rest is the world's remaining forests.

Let's put this into context: Demand for food will at least double by 2050, increasing demand for more land.

So no wonder, then, that there's a remarkable land grab under way.

Since 2000, there have been tens of thousands of land deals involving governments, large businesses, a range of "conglomerates," hedge funds, and vaguely defined institutions based in countries such as China, Saudi Arabia, Qatar, Norway, France, the UK, Germany, Indonesia, and the United States.

These organizations are buying up very large amounts of land in various regions of the world, including sub-Saharan Africa, Asia, and South America. They are doing so in order to log timber; or mine metals, minerals, rare earths, and phosphates; or simply to clear land for livestock or crops.

Mirny, Russia

During just the past twelve years, almost 125 million acres of land have been traded. That is an area of land approaching half the size of Western Europe being bought and sold to foreign governments and businesses.

But that's not the most important story.

Land use, land degradation, loss of habitat, and pollution runoff are now causing significant species population loss.

The International Union for the Conservation of Nature (IUCN)—the world's leading authority on biodiversity—estimates that, as of 2012, almost 31 percent of all amphibians, 21 percent of all mammals, and 13 percent of all birds are threatened with extinction.

We are now almost certainly losing species at a rate up to one thousand times faster than we would expect from ordinary "background" (natural) processes.

This means that human activity is almost certainly now set to cause the greatest mass extinction of life on Earth since the event that wiped out the dinosaurs 65 million years ago.

When newspapers, TV, and green campaigns want to highlight loss of species, they often do so by showing a picture of a lonely looking polar bear on a minuscule ice floe, looking as though "this is it."

But losing polar bears is just the tip of the iceberg, no pun intended. What we need to be a lot more concerned about is the loss of biodiversity itself.

Species Extinction

Extinction rate

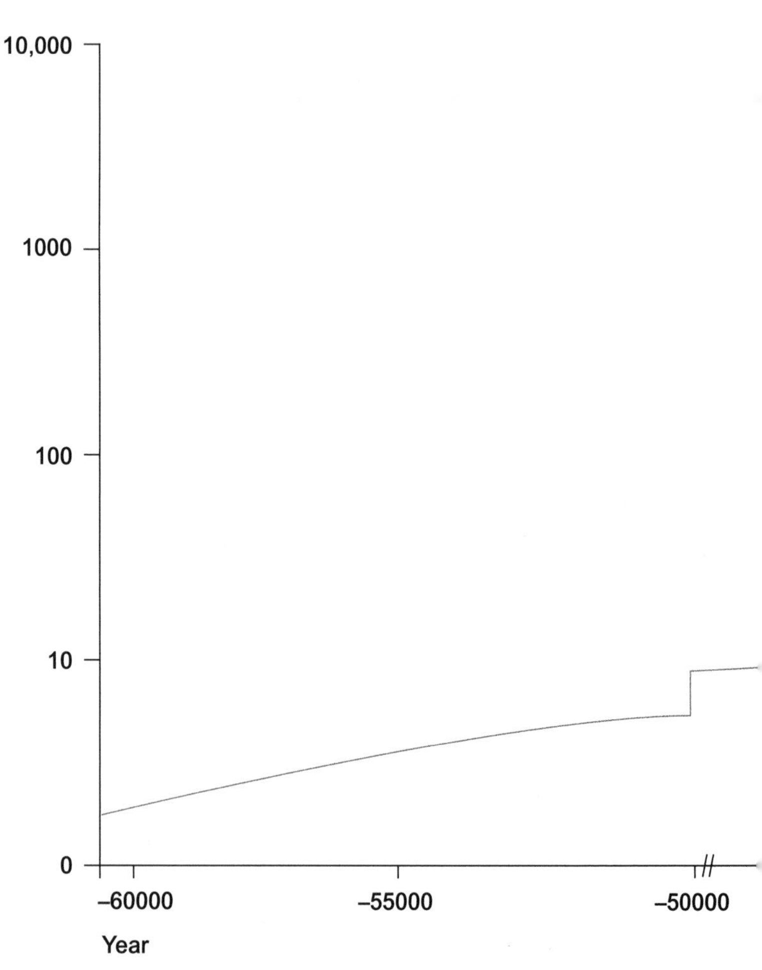

We are embarking upon the sixth mass extinction of life on Earth. Extinction rates are expressed as number of extinctions per million species-years (E/MSY) from the Pleistocene to 2050 (past and present rates from the Paleobiology Database [PBDB 2010]). Extinction rate increases to 9 E/MSY in Pleistocene during the mass extinction of large megafauna from humans expanding out of Africa. Extinction rate increases during the Holocene to 24 E/

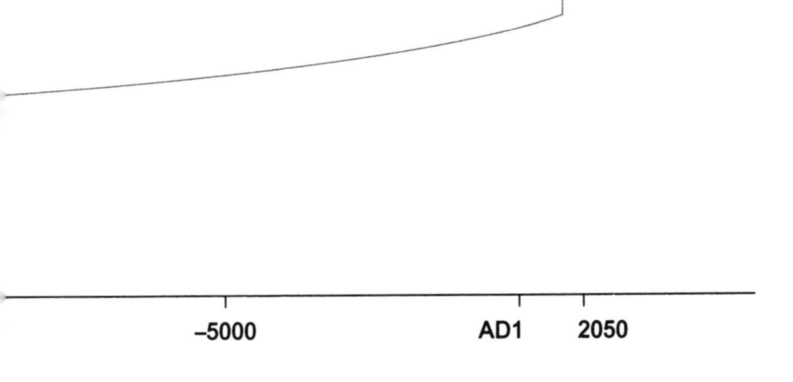

MSY. A rapid rise in extinction rates then occurred from 1800 (the sharp jump in extinction rate observable on the graph) and is predicted to continue, principally from loss of habitat. Data based on S. Pimm & P. Raven, "Biodiversity: Extinction by numbers," *Nature,* 403 (2000); A. Barnosky, et al., "Has the Earth's sixth mass extinction already arrived?" *Nature,* 471 (2011)

Loss of Tropical Rainforest and Woodland

Percentage of 1700 value

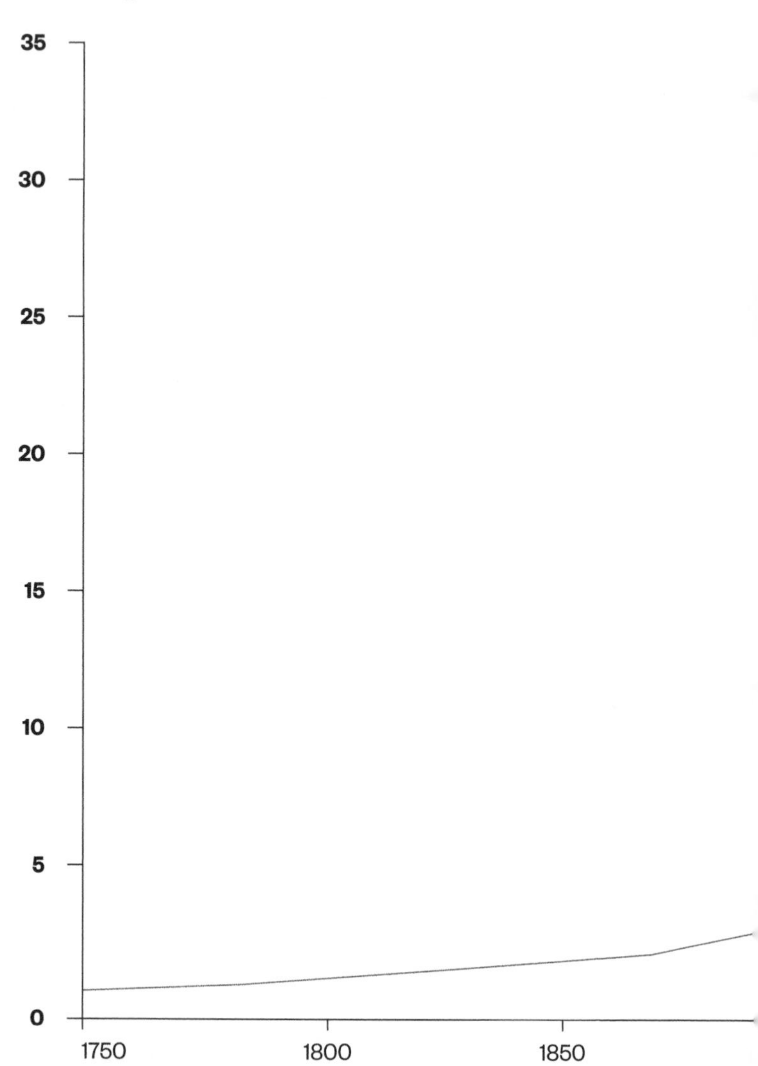

The rate at which we are losing tropical rainforests and woodlands is increasing every year, and has done so since at least 1700. (Here expressed as a percentage of the 1700 value.) "Plans" to halt this loss have failed,

and continue to fail. Adapted from Steffen, et al., "The Anthropocene: From Global Change to Planetary Stewardship," *AMBIO*, October 2011 (Royal Swedish Academy of Sciences)

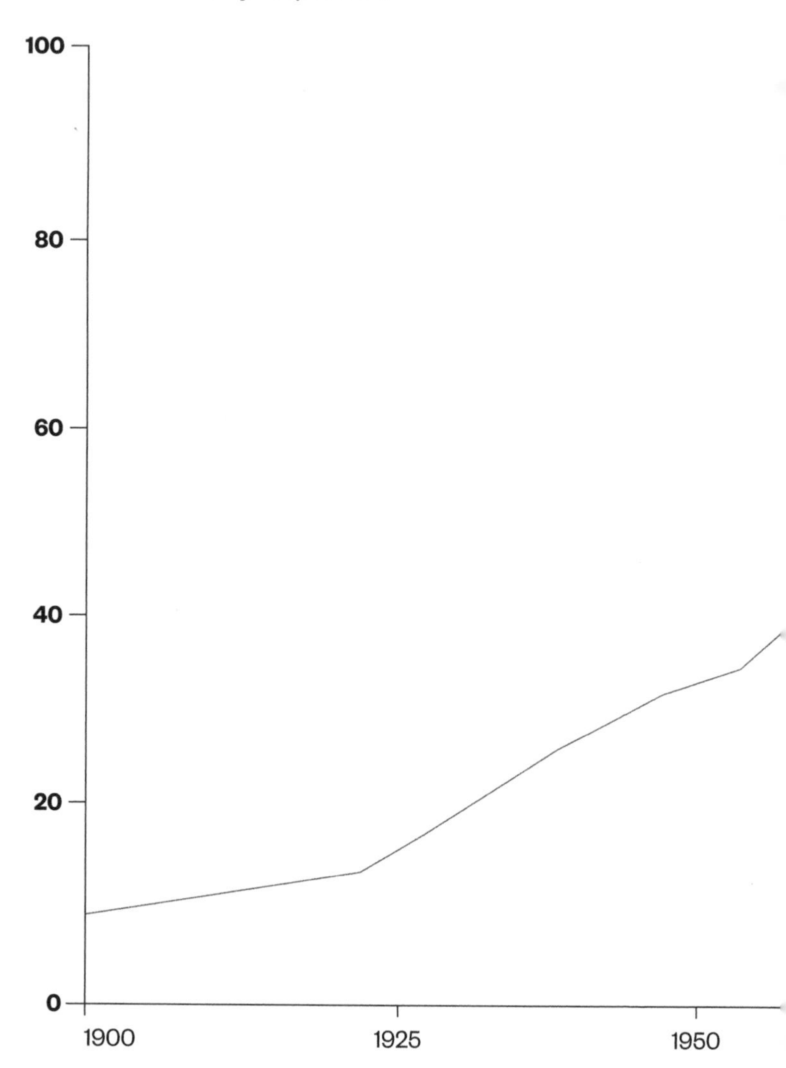

Since 1900, the percentage of the world's oceans either fully exploited or overexploited has risen from less than 10 percent to possibly as much as 87 percent. We are harvesting ocean ecosystems at a rate that is completely unsustainable. Adapted from Steffen, et al., "The Anthro-

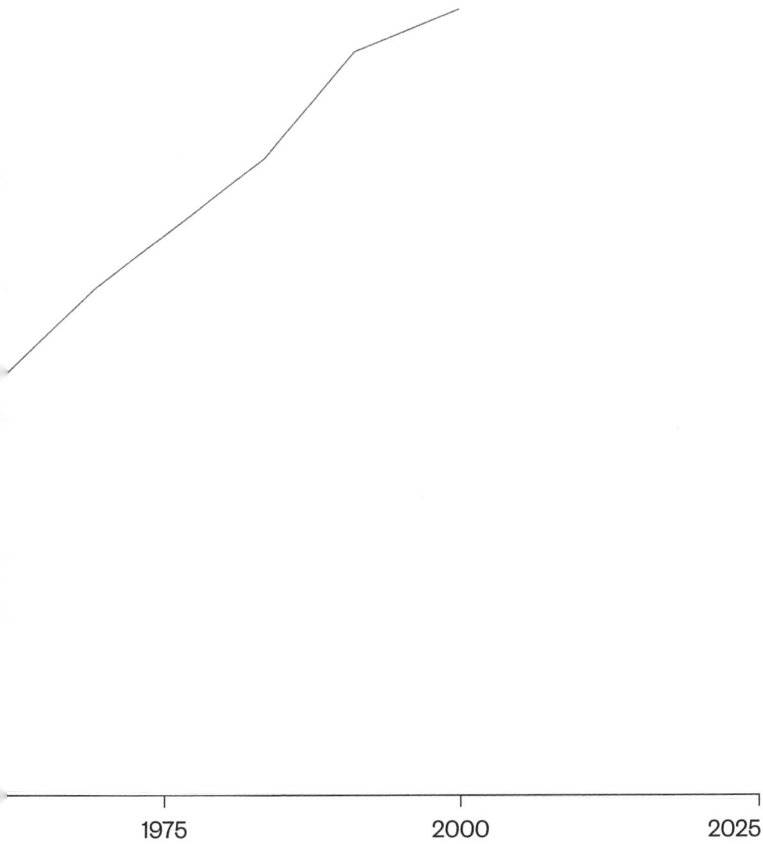

1975 2000 2025

pocene: From Global Change to Planetary Stewardship," *AMBIO*, October 2011 (Royal Swedish Academy of Sciences); FAO, *The State of the World's Fisheries 2012* (2012); B. Worm, et al., "Rebuilding global fisheries," *Science*, 325 (2009)

Because it turns out that biological diversity is not just a "nice thing to have."

It is the very diversity of life on Earth—the diversity that we are rapidly eroding—that produces and provides many vital planetary functions, including numerous "ecosystem services," which is to say the things nature provides "for free." Like our water, our food, and our climate.

Indeed, the loss of biodiversity on the current scale is inevitably going to mean loss of ecosystem function, and loss of ecosystem function is going to mean loss of vital ecosystem services.

The loss of ecosystem services poses a very real threat to our survival.

Let's address the issue of our food.

That food demand is increasing is not surprising. What is surprising is that food demand is accelerating at a far faster rate than population growth.

Why? There are three reasons.

The first is that more people are eating more food. As GDP (gross domestic product, the standard measure of the wealth of a country and its inhabitants) increases, calorie consumption also increases. As we get richer (or suffer less poverty), we consume more food.

The second reason is that more of us are not only consuming more, we're consuming differently. In particular, a rapidly growing number of people in developing regions such as Brazil, Africa, and China are eating more meat.

Third, for hundreds of millions of consumers, eating has now become a recreation—a pastime.

This is further increasing pressure on both food production and land use. More meat consumption is turning out to mean more soy production. Soy is a principal animal feed, used because it is a fast way to produce animal protein. The rapidly expanding use of land for soy production to feed livestock, plus the rapidly expanding use of land to keep livestock on, is putting significant additional pressure on land use and deforestation.

Soy plantation, Vilhena, Brazil

Critically, the entire global food production system depends entirely upon a stable climate. But right now, the climate is anything but stable. And it is set to become more and more unstable.

And think about this: Producing more food itself is going to accelerate climate change.

Food production accounts for around 30 percent of all greenhouse gases produced by human activity from carbon dioxide, methane, and nitrous oxide (a greenhouse gas three hundred times more potent than CO_2, generated as a by-product of our fertilizer use). That is more than manufacturing or transportation.

More food is going to mean more greenhouse gas emissions, accelerating climate change, and extreme weather events, increasingly threatening the future of our food.

We can already see worrying signs of things to come. The 2008 Australian drought, the 2010 Russian and Eastern European droughts, and the 2012 drought in the United States—the world's largest grain producer—as a consequence of extreme weather and record heat waves—led to losses of between 20 and 40 percent of the entire grain and corn harvest each year.

These losses, in turn, pushed up basic food prices steeply on the futures market. Which made basic foodstuffs much more expensive, especially for the poor.

Right now, over one billion people are living in conditions of extreme water shortage.

Yet our consumption of water is accelerating rapidly.

A staggering 70 percent of Earth's available fresh water is used for the irrigation of agriculture.

Much of this water comes from underground water supplies called "aquifers." These are now being depleted faster—much faster—than they can be replenished. Yet we are going to have to increase irrigation significantly this century.

Our water use is increasing rapidly in other ways, too. Take one important, yet little-known aspect of increasing water use: "hidden water."

Hidden water is water used to produce things we consume but typically do not think of as containing water. Such things include chicken, beef, cotton, cars, chocolate, and mobile phones.

For example: it takes almost 800 gallons of water to produce a burger. In 2012, about five billion burgers were consumed in the UK alone. That's about 4 trillion gallons of water—on burgers. Just in the UK. It's estimated that at least 21 billion burgers are consumed in the U.S. each year. That's more than 16 trillion gallons of hidden water to produce burgers in the United States, in one year.

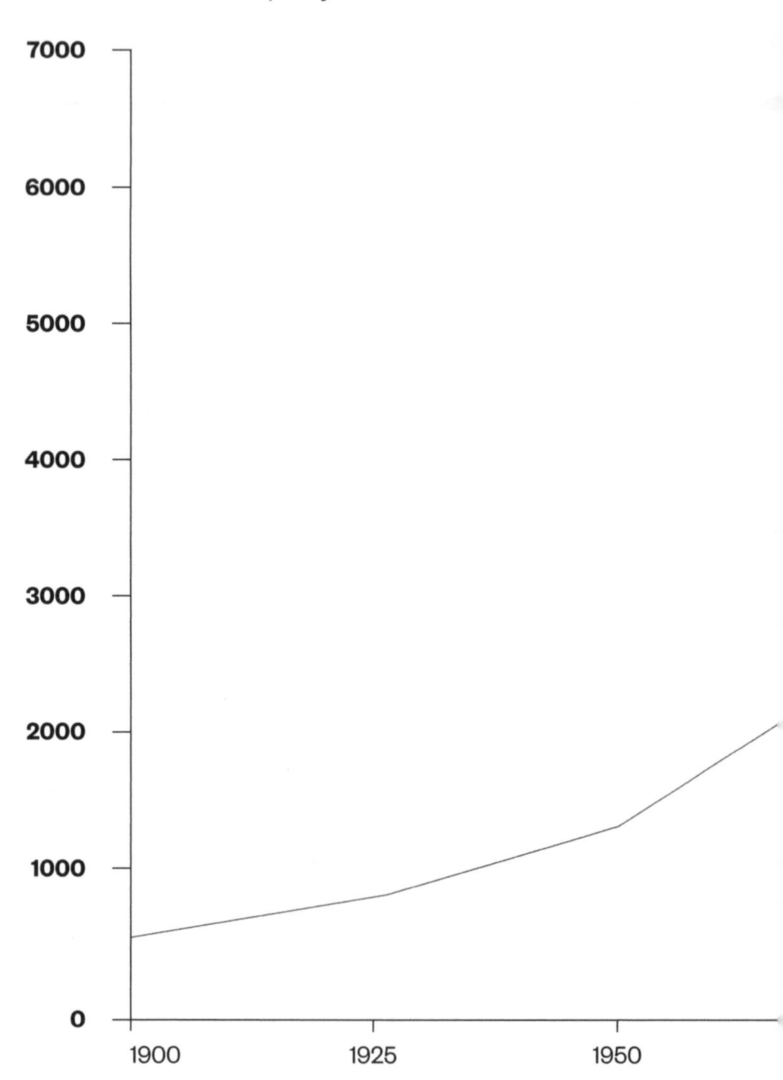

A hundred years ago our water use was around 600 cubic km/year. It is currently (conservatively) 4,000 cubic km/yr. By 2025 we are likely to be using at least 6,000 cubic km/year of water globally. Water use is growing at double the

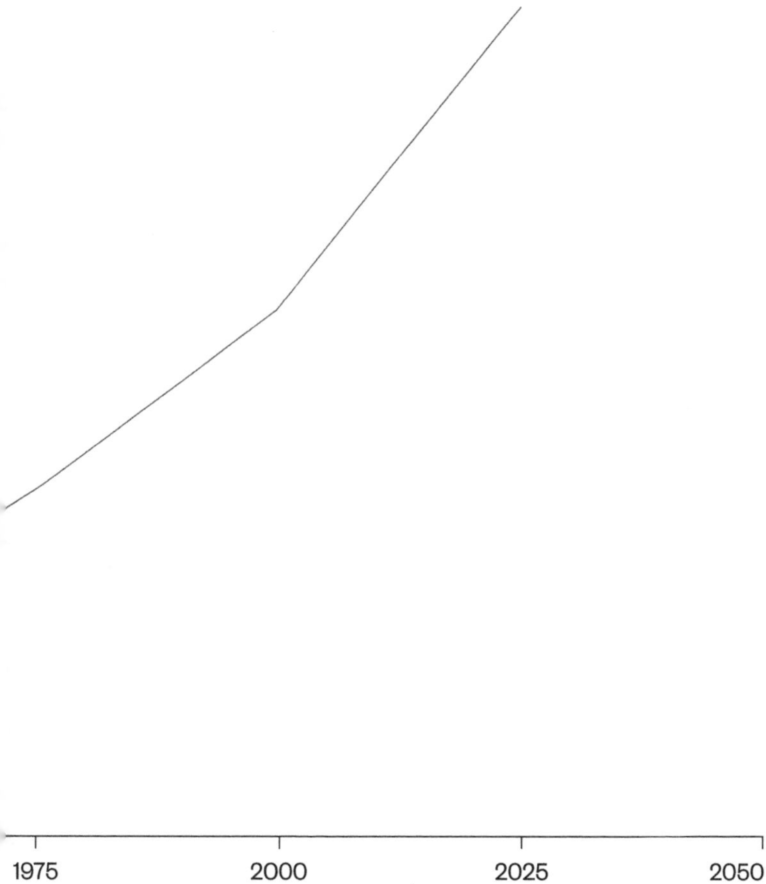

1975 2000 2025 2050

rate of population growth. (Data and projections are water extraction.) Adapted from UN Environment Programme Water Statistics, 2008; FAO

It takes about 2,400 gallons of water to produce a chicken. In the U.S. alone we consume around 8 billion chickens a year.

It takes about 7,000 gallons of water to produce one kilogram of chocolate. That's roughly 304 gallons of water per Hershey bar. This should surely be something to think about while you're curled up on the sofa eating it in your pajamas.

But I have bad news about pajamas. Because I'm afraid your cotton pajamas take 2,300 gallons of water to produce.

It takes 26 gallons of water to produce a cup of coffee. And that's before any water has actually been added to your coffee. Worldwide, two and a quarter billion cups of coffee are consumed every day.

And—irony of ironies—it takes something like four liters of water to produce a one-liter plastic bottle of water. In 2011, Americans consumed per capita 222 bottles of water—that's approximately 70 billion bottles of water in a year for the whole country. Last year, in the UK alone, we bought, drank, and threw away nine billion plastic water bottles. That is 36 billion liters of water, used completely unnecessarily. Water wasted to produce bottles—for water.

It takes about 19 gallons of water to produce one of the "chips" that typically powers your laptop, GPS, phone, iPad, TV, microwave, camera, and your car. There were probably something like 3 billion such chips produced in 2012. That is at least 57 billion gallons of water. On semiconductor chips.

In short, we're consuming water, like food, at a rate that is completely unsustainable.

The term "peak oil" is an increasingly familiar one.

It refers to the point at which maximum rate of oil extraction is reached, beyond which it starts to decline. The generally accepted claim is that we've reached peak oil—and that we're heading for a global energy crisis soon, as we run out of oil and gas.

But it is almost certainly not true. There are enormous reserves of "proven" oil and gas. And every year, we are discovering significant new oil and gas deposits, from Brazil to the Arctic. On top of that there is the so-called "energy game-changing revolution" in the U.S. that is shale oil and gas.

So I'm not worried about running out of oil and gas. I'm worried that we're going to continue using them. Doing so will simply accelerate the climate problem even further.

But guess what? This is exactly what's happening. In 2012, the U.S. energy company Exxon—the world's largest oil producer—signed a deal with Russia to invest up to $500 billion in oil and gas exploration and extraction in the Arctic, in Russia's Kara Sea.

Why? Because climate change is already making oil and gas exploration economically viable: the Kara Sea is no longer covered in thick ice all year round.

And President Barack Obama has committed to extending the import of tar sands oil from Alberta in Canada to the U.S. through the development of the "Keystone XL" project—providing U.S. consumers with close to 1 million barrels of oil per day from Canadian tar sands.

In the UK, despite its stated commitment to tackling climate change, the British government issued 167 new licenses to drill for oil and gas in the North Sea—the largest number since North Sea oil drilling began in 1964.

John Hayes, then the UK energy minister, described this as "A great week for the oil and gas industry. There's this myth that the North Sea has seen its best days, but this shows there is still a lot of opportunity. The Government is taking the right action to offer certainty and confidence to investors."

And we are not just increasing our consumption of oil and gas globally. We are increasing our use of coal. Even the UK increased its consumption of coal for energy production by 31 percent in 2012.

While our politicians, our businesses, and our own stupidity look set to ensure we remain lethally addicted to oil, gas, and coal, it is worth pointing out that hundreds of millions of people rely, every day, on burning wood just to get by.

Indeed, the use of wood for cooking is now a significant source of deforestation in parts of Africa. Such is the use of wood and charcoal for cooking throughout Africa and Asia that it is producing unprecedented amounts of what is called "black carbon"—principally soot. More black carbon is now produced every year than in the whole Middle Ages.

It is a major problem in many developing countries. It is contributing significantly to both short-term climate variations and long-term climate change.

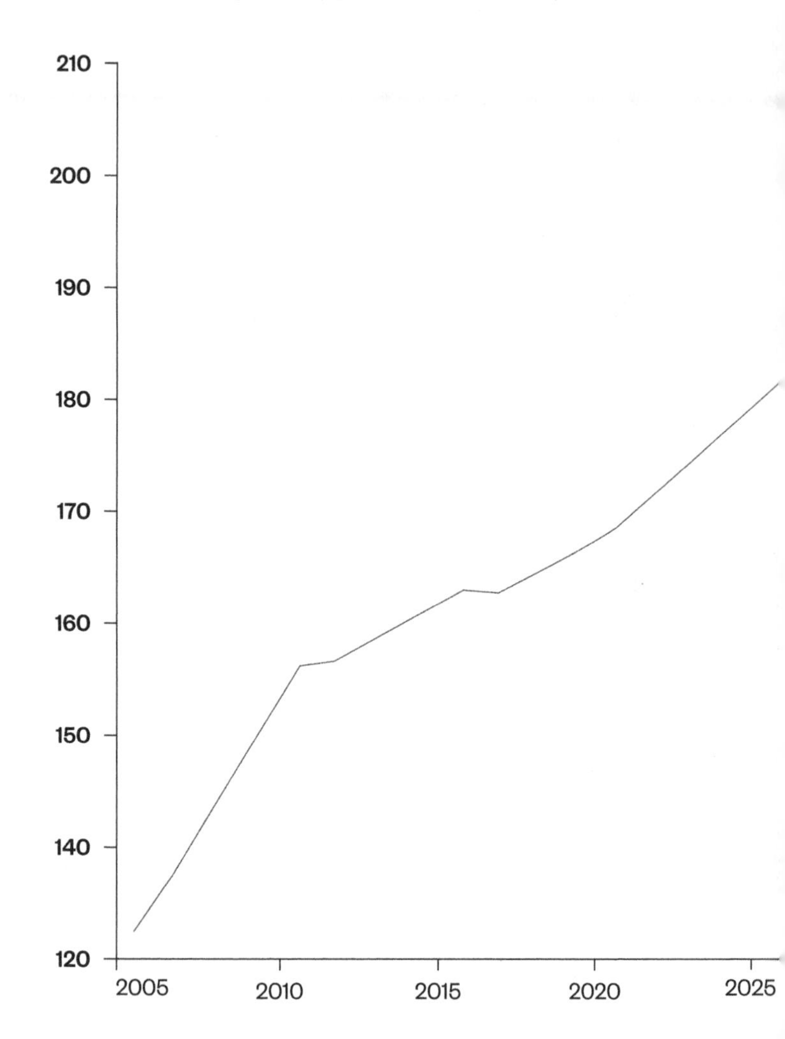

Our use of coal—one of the world's major causes of CO_2 emissions—is increasing every year to manufacture and power all the stuff we're consuming. U.S. coal exports to China doubled between 2011 and 2012, to power the factories that make stuff Americans want to consume, which are

2030 2035 2040 2045 2050

then exported back to the United States. The U.S. is exporting its CO_2 emissions. We are set to remain addicted to coal, oil, and gas. Adapted from U.S. Energy Information Administration, *International Energy Outlook 2011* (EIA, 2011)

An atmospheric brown cloud shrouds Hong Kong

But black carbon doesn't just come from poor people in poor countries burning wood and charcoal just to get by every day. It also comes from rich people (like us), in rich countries (like the U.S., UK, Germany, Canada, and Australia) transporting ourselves, and our stuff, via planes, ships, and cars.

Combined, the black carbon produced by developing countries (from burning wood) and from developed countries (from cars, planes, ships, and manufacturing) creates what are known as atmospheric brown clouds (ABCs).

ABCs are having really very significant effects on human health, in terms of respiratory diseases and premature deaths. The polluting effects of ABCs affect some three billion people worldwide.

The total number of motor vehicles produced since 1900 exceeded 2 billion in 2012.

And let's put that into context: if current demand continues, some 4 billion more cars could be produced in the next forty years.

VW lot, Texas

What does a car cost?

Volkswagen, Ford, Toyota, and others keep telling us that we can buy a car from around $13,000.

That's not what a car costs. Let's look at what a car costs.

The iron ore forming the basis of the car's steel body has to be mined (from somewhere like Australia). It is then transported on a very large and very polluting ship to somewhere like Indonesia or Brazil, to be made into steel.

That steel is then transported on a very large and very polluting ship to a car factory in, say, Germany.

Used cars

The tires have to be manufactured. The rubber has to be produced in Malaysia, Thailand, or Indonesia. The rubber then has to be shipped to a country that manufactures tires.

The plastic for the car dashboard starts out as oil in the ground. That oil has to be extracted, and exported—on a very large and very polluting ship—to be made into plastic, which then gets transported to the car factory to then be molded into a dashboard.

The leather for the seats came from an animal. The animals needed to produce the seats—cows—require a lot of water and a lot of food. They will have been reared somewhere such as Brazil. Their skins will have been shipped to somewhere such as India for processing. (Kanpur is the center of India's booming leather industry, producing leather for car seats and handbags for the U.S., UK, and Europe. The leather processing factories pollute both the atmosphere and the river Ganges with hydrochloric acid, chromium, and a cocktail of other poisonous chemicals.) The resulting processed leather will then be shipped to a car factory to be made into seat covers.

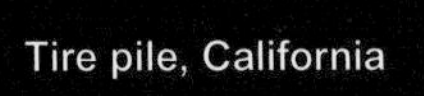

Tire pile, California

The lead in the battery has to be mined in China, for example, then shipped and made into batteries. Batteries that need to be transported on a very large and very polluting ship to car factories in Germany, the United States, and elsewhere.

All this before a single car is even assembled. Let alone before a car is then transported for you to buy.

And that's before you've put a single gallon of gas in your car and started contributing further to the climate problem.

What is the cost of a car?
An absolute fortune.

But you don't have to pay the *real* cost—that is to say, the cost of environmental degradation; pollution from mining, industrial processes, and transportation; the resultant loss of ecosystems; and climate change. What economists like to call "externalities."

At least not yet. But this cost—the cost of the consequences of producing a car, the *real* cost of producing a car—will have to be paid for by someone in the future.

Maybe you. More likely your children.

Transportation

Number of motor vehicles (millions)

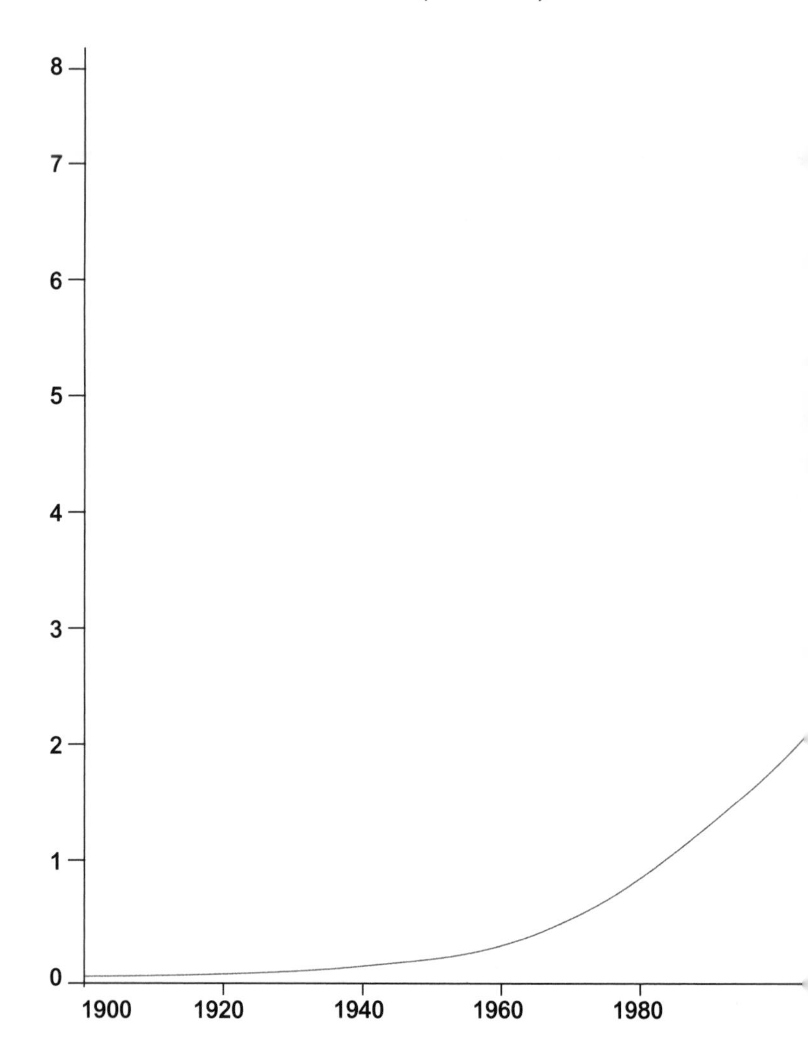

Between 1900 and 2012, some 2.6 billion automobiles (cars, buses, trucks) were produced. By 2050, there could be an additional 4 billion new automobiles produced (projection is based upon an annual growth rate of 2.5 percent. The average rate of growth from 1998–2011 was 3.27 percent). Data

2020 2040 2060 2080 2100

and projections based on data from the International Organization of Motor Vehicle Manufacturers (Organisation Internationale des Constructeurs d'Automobiles, OICA), Dargay et al. (2007), "Vehicle ownership and income growth, worldwide: 1960–2030" and calculations by the author.

This year we will fly four trillion miles.

When you consider that the average long-haul flight uses around 110 tons of fuel, that amounts to a lot of pollution and greenhouse gas emissions.

It's worth reminding ourselves that our stuff does not come from Target, Whole Foods, A&P, Amazon, Walmart, or Best Buy. Our stuff comes from countries like China, Morocco, Brazil, Turkey, Spain, South Korea, and Peru. Whether it's asparagus, pajamas, or electronics.

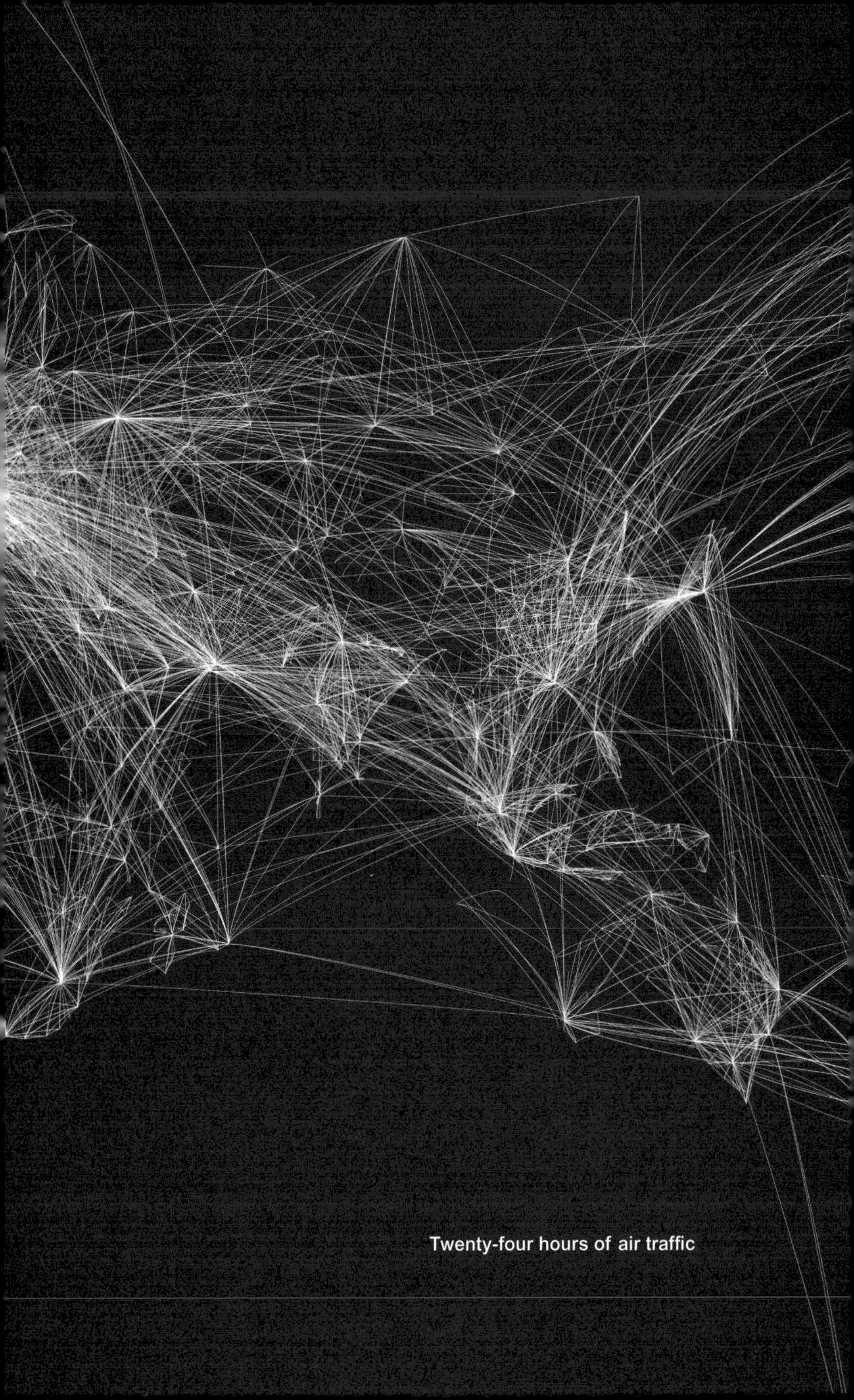

Twenty-four hours of air traffic

Something like 500 million containers of stuff—stuff that we will consume—from Japanese and German cars, South African oranges, Peruvian asparagus, and Kenyan cut flowers, to T-shirts and dresses from Morocco or Vietnam, sneakers, music players, laptops, mobile phones, and televisions from China or South Korea—will be handled and transported around the world this year. Plus billions of tons of raw materials that will form the basis of our consumption—metals, phosphates, grain, oil, gas, and coal.

Right now, climate change is accelerating.

We should more properly say an accelerating change in the Earth System—of which our climate is one component.

This impact is most obvious at high latitudes. What this means is that if we want to see what our future looks like, the Arctic is the place to look first. And it doesn't look good.

Arctic coastlines are retreating by up to 30 meters per year in areas such as the Laptev Sea and Beaufort Sea. Greenland and Antarctica are now losing somewhere between 300 billion and 600 billion tons of ice mass per year into the sea.

And to make matters worse, probably much worse, melting sea ice caused by our activities is now causing the release of significant quantities of methane from the Arctic Ocean.

For the first time, over a hundred plumes of methane—many of them over half a mile in diameter—have been observed rising from previously frozen methane stores in the East Siberian Sea. Indeed a conclusion was that thousands of such plumes, many of them nearly a mile across, now exist.

This could be very big trouble on a very big scale.

Methane is many times more potent a greenhouse gas than CO_2. If, as seems likely, melting sea ice, triggered by our activities, is now causing the release of this methane, it will go on for decades, possibly centuries, and we will be completely unable to stop it.

Almost all of the data that's emerging now from the Arctic is worse—far worse—than the most extreme predictions of even ten years ago.

But of course it's not just the Arctic. It's everywhere.

Think about this: Right now, every leaf on every tree on Earth is experiencing a level of CO_2 that the planet has not experienced for millions of years.

How the planet's plants will respond to this we simply don't fully understand.

This is critically important, because the planet's plants (principally forests) are fundamental components of what's called the "global carbon cycle."

The global carbon cycle is the vast and highly complex system that processes the earth's carbon—hundreds of billions of tons of it every year. Importantly, the global carbon cycle currently does all seven billion of us an enormous favor in terms of slowing down climate change. This is because the planet's plants and oceans are absorbing about 50 percent of all of our CO_2 emissions (and have been since the industrial revolution). The other 50 percent of the CO_2 we produce remains in the atmosphere. This is the CO_2 that is predominantly responsible for climate change.

But this favor may be about to come to an end.

One of the consequences of increasing human activity—for example, deforestation and human-induced climate change—could be that the global carbon cycle will switch from being a net carbon "sink" (absorber) to a net carbon source (producer), thus further accelerating climate change.

The global carbon cycle is absolutely critical to the ability of us, and almost every other species on Earth, to survive on this planet.

We are now radically changing every component of the global carbon cycle.

Even *The Economist* has pointed out that the cost of climate change, the cost of resource exploitation (mining for metals, extraction of oil and gas), especially in the Arctic, which itself will further accelerate climate change, is almost certainly going to be unimaginable damage to whole ecosystems, with unimaginable costs in terms of our future food, water, and ecosystem services that are simply too difficult to contemplate. *The Economist*'s conclusion (I paraphrase): No one wants to be responsible unilaterally, but all are happy to profit from it. This is a textbook case of the commons despoiling tragedy that is climate change.

All of the science points to the inescapable fact that we are in trouble. Serious trouble.

And right now, we are heading into completely uncharted territory as our population continues to grow toward ten billion.

But one thing that is predictable is that things are going to get worse.

What kinds of challenges do we face over the coming years, as a consequence of our growing population and our activities?

The land problem is simple: We are already using 40 percent of all available land on Earth for food production. Yet let's remind ourselves that demand for food is going to double—at least—by 2050.

This means that pressure to clear many of the world's remaining tropical forests—rain forests—for human use is going to intensify every decade. Because this is predominantly the only available land that is left for expanding agriculture at scale. Unless Siberia thaws out before we finish deforestation.

By 2050, some 2.5 trillion acres of land are likely to be cleared to meet rising food demands from a growing population. This is an area greater than the size of the United States. And accompanying this will be 3.3 billion tons per year of extra CO_2 emissions that will result.

If Siberia does thaw out before we finish our deforestation of much of the remaining tropical forest, it would result in a vast amount of new land being available for agriculture, as well as opening up a very rich source of minerals, metals, oil, and gas. In the process this would almost certainly completely change global geopolitics. Siberia's thawing would turn Russia into a remarkable economic and political force this century because of its newly uncovered mineral, agricultural, and energy resources.

It would also inevitably be accompanied by vast stores of methane—currently sealed under the Siberian permafrost tundra—being released, accelerating our climate problem even further.

Meanwhile, another three billion people are going to need somewhere to live.

By 2050, 70 percent of us are going to be living in cities. This century will see the rapid expansion of cities, as well as the emergence of entirely new cities that do not yet exist. It's worth mentioning that of the nineteen Brazilian cities that have doubled in population in the last ten years, ten are in the Amazon. All of this is going to use yet more land.

The food problem is also simple.

We currently have no known means of being able to feed ten billion of us at our current rate of consumption and with our current agricultural system.

Indeed, simply to feed ourselves in the next forty years, we will need to produce more food than the entire agricultural output of the past 10,000 years combined.

Yet food productivity is set to decline, possibly very sharply, over the coming decades.

Why? There are three reasons.

First, climate change. Climate change is going to increase the frequency and severity of extreme weather events (it is already producing unusual heat waves, droughts, floods) resulting in increasing loss of crops in many parts of the world.

The second is soil degradation and desertification, both of which are increasing rapidly in many parts of the world. They are the result of water runoff, pollution (including fertilizers and salination from irrigation), intensified agricultural practices, and overgrazing.

The third is water stress, deriving from more frequent and severe climate change–induced droughts, rapidly increasing hidden and other water use by a growing population, and the rise in the consumption of just about everything.

If we want to get just a glimpse of what we can expect this year or next year, certainly over the decades to come, we need only look again at the impact of recent heat waves in Australia (2008), Russia (2010), and the United States (2012), which destroyed up to 40 percent of grain and corn harvests, and in which livestock died in the tens of thousands.

In the heat wave of 2010, the Russian government placed an embargo on grain exports, which caused chaos in the commodities markets, an unprecedented food price spike, and, consequently, food riots across Asia and Africa—unrest that led to the violence of what we now refer to as the "Arab Spring."

But there are in addition two other emerging crises that threaten the future of our food. The first is a phosphate crisis. The amount of food we produce is almost entirely dependent upon phosphate-based fertilizers. But phosphate reserves are finite, and it is becoming apparent that we are going to run out of it, almost certainly sometime this century. The question is "when," not "if." And because we are so heavily reliant on phosphates, when that happens it is the end of food as we know it for the global human population.

Food riots, Algeria, 2011

The second is the emergence of novel fungal pathogens that threaten to devastate crops (and, potentially, livestock). Indeed, even for all the major known fungal pathogens, only a class of chemicals called triazoles are still effective in combating fungal disease in crops (but fungal pathogens are rapidly evolving resistance to them). If, or when, this happens, we are in danger of famine-like starvation scenarios on a large scale.

The water problem is this:

By the end of this century, large parts of the planet will not have anywhere near enough usable water.

Billions of people are likely to be living in conditions of extreme water shortage as a result of increasing climate change, increasing food demand, and an increasing population.

First of all, unprecedented, large-scale changes to the global "hydrological cycle" (the planet's water cycle), which are already well under way as a result of human activity and human-induced climate change and are set to significantly increase this century. These changes are set to have a very significant negative impact on water availability.

Second, our use of groundwater (essential for irrigating our food) is accelerating rapidly, far faster than groundwater is or can be replenished. We face a highly dangerous and accelerating shortage of groundwater. And efforts to "technologize" our way out of this problem through approaches such as water diversion, artificial groundwater recharge, and "efficient" irrigation technologies have all failed.

Moreover, freshwater supplies stored in the planet's glaciers and snow cover are projected to decline alarmingly this century.

And even small rises in water temperature, plus increasing extreme weather events causing both drought and flooding, together with increasing pollution of water supplies through fertilizer, metals, and industrial agents, are projected to significantly affect water quality (making water unusable), and significantly exacerbate numerous forms of water pollution.

I'm afraid our water problem is unavoidably going to have very adverse consequences for agriculture, human health, and ecosystems.

You might not be surprised to learn that global car production, air traffic, and shipping are all set to increase significantly this century.

For starters, we are set to produce at least three times more cars this century than we did in the twentieth century. And the global shipping and airline sectors are projected to continue to expand rapidly every year, year on year, transporting more of us, and more of the stuff we want to consume, around the planet.

That is going to cause enormous problems for us in terms of more emissions, more black carbon, and more pollution from mining and processing to make all this stuff.

But think about this. In transporting us and our stuff all over the planet, we are creating a highly efficient network for the global spread of potentially catastrophic diseases.

There was a global pandemic just ninety-five years ago—the Spanish flu pandemic, which is now estimated to have killed up to 100 million people. And that's before one of our more questionable innovations—budget airlines—were invented.

The combination of millions of people traveling around the world every day, plus millions more people living in extremely close proximity to pigs and poultry—often in the same room, making a new virus jumping the species barrier more likely—means we are increasing, significantly, the probability of a new global pandemic.

So no wonder then that epidemiologists increasingly agree that a new global pandemic is now a matter of "when" not "if."

Our energy problem is simple.

We are going to have to triple—at least—energy production by the end of this century to meet expected demand.

To achieve this, we will need to build, roughly speaking, something like:

1,800 of the world's largest dams;

23,000 nuclear power stations;

14 million wind turbines;

36 billion solar panels;

or just keep going with predominantly oil, coal, and gas—and build the 36,000 new power stations that means we will need.

Our existing oil, coal, and gas reserves alone are worth trillions of dollars. Are governments and the world's major oil, coal, and gas companies—some of the most influential corporations on Earth—really going to decide to leave this money in the ground, as demand for energy increases relentlessly?

I doubt it.

The emerging climate problem is on an entirely different scale.

The problem is that we may well be heading toward a number of critical "tipping points" in the global climate system.

All complex systems, such as the earth's system, are characterized by one important feature: a very small change ("perturbation") can lead to an extraordinarily large and unpredictable impact that "tips" the system into an entirely different and unpredictable state.

Let's take just one of the tipping points we're heading for: a rise in global average temperature of above 2 degrees Celsius.

There is a politically agreed global target—driven by the Intergovernmental Panel on Climate Change (IPCC)—to limit the global average temperature rise to 2 degrees Celsius. The rationale for this target is that a rise above 2 degrees carries a significant risk of catastrophic climate change that would almost certainly lead to irreversible planetary "tipping points" caused by events such as the melting of the Greenland Ice Shelf, the release of frozen methane deposits from Arctic permafrost, or dieback of the Amazon.

But in fact the first two are happening now—at below the 2-degree threshold. As for the third, we're not waiting for climate change to do this—we're doing it right now through deforestation.

And unfortunately, recent research shows that we look certain to be heading for a larger rise in global average temperature than 2 degrees—a far larger rise.

It is now very likely that we are looking at a future global average temperature rise of 4 degrees—and we can't rule out a rise of 6 degrees.

A 4- to 6-degree rise in global average temperature will be absolutely catastrophic. It will lead to runaway climate change, capable of tipping the planet into an entirely different state, rapidly. Earth would become a hellhole.

In the decades along the way, we will witness unprecedented extremes in weather, fires, floods, heat waves, loss of crops and forests, water stress, and catastrophic sea-level rises.

But even if we're lucky enough to fall short of anything like a 4- to 6-degree rise in global temperature, there almost certainly won't be a country called Bangladesh by the end of this century—it will be underwater.

Large parts of Africa will become permanent disaster areas. The Amazon could be turned into savannah or even desert. And the entire agricultural system will be faced with an unprecedented threat.

More “fortunate” countries such as the United States, the UK, and most of Europe may well look like something approaching militarized countries, with heavily defended border controls designed to prevent millions of people who are on the move from entering, because their own country is no longer habitable, or has insufficient water or food, or is experiencing conflict over increasingly scarce natural resources.

These people will be “climate migrants.” The term “climate migrants” is one we will increasingly have to get used to.

Indeed, anyone who thinks that the emerging global state of affairs does not have great potential for civil and international conflict is deluding themselves.

It is no coincidence that almost every scientific conference that I go to about climate change now has a new type of attendee: the military.

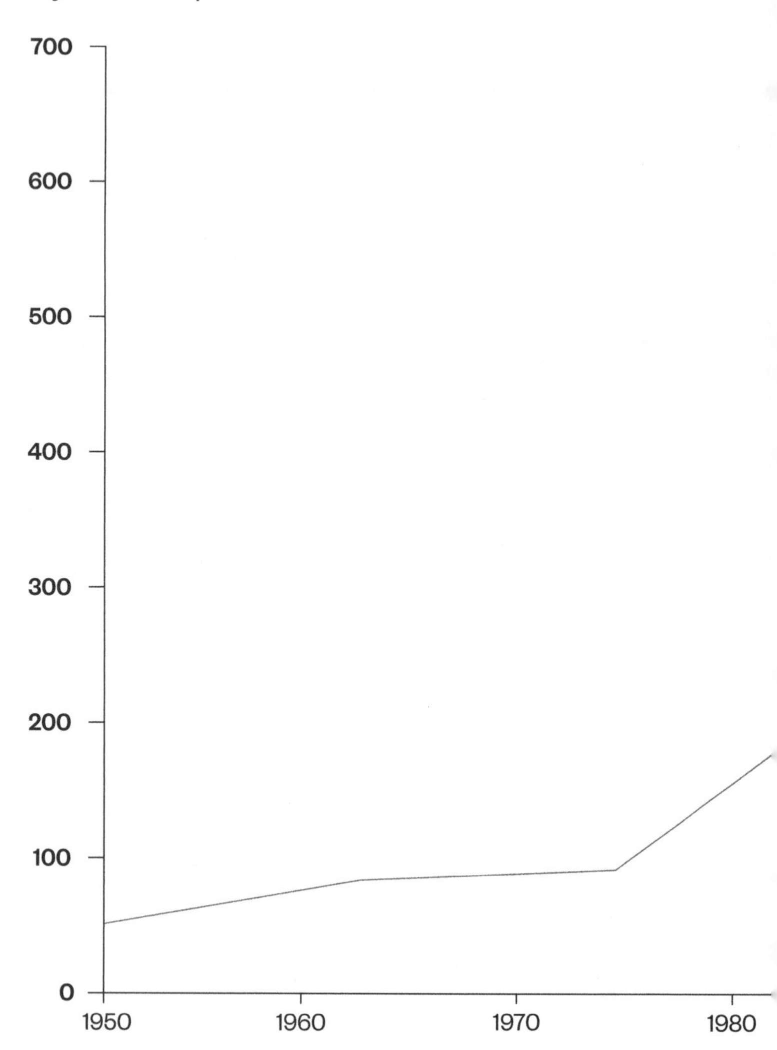

Extreme weather events are increasing rapidly. Since 1950, the number of major floods in Asia has increased from fifty per decade to nearly seven hundred per decade. To deny that this is almost certainly because of

1990 2000 2010 2020

human-induced (anthropogenic) changes to the climate and the entire Earth System would be foolish. Adapted from ICIMOD & UNEP-Grid Arendal, 2010, and Millennium Ecosystem Assessment Report, 2000.

Fires in the Americas

Major fires per decade

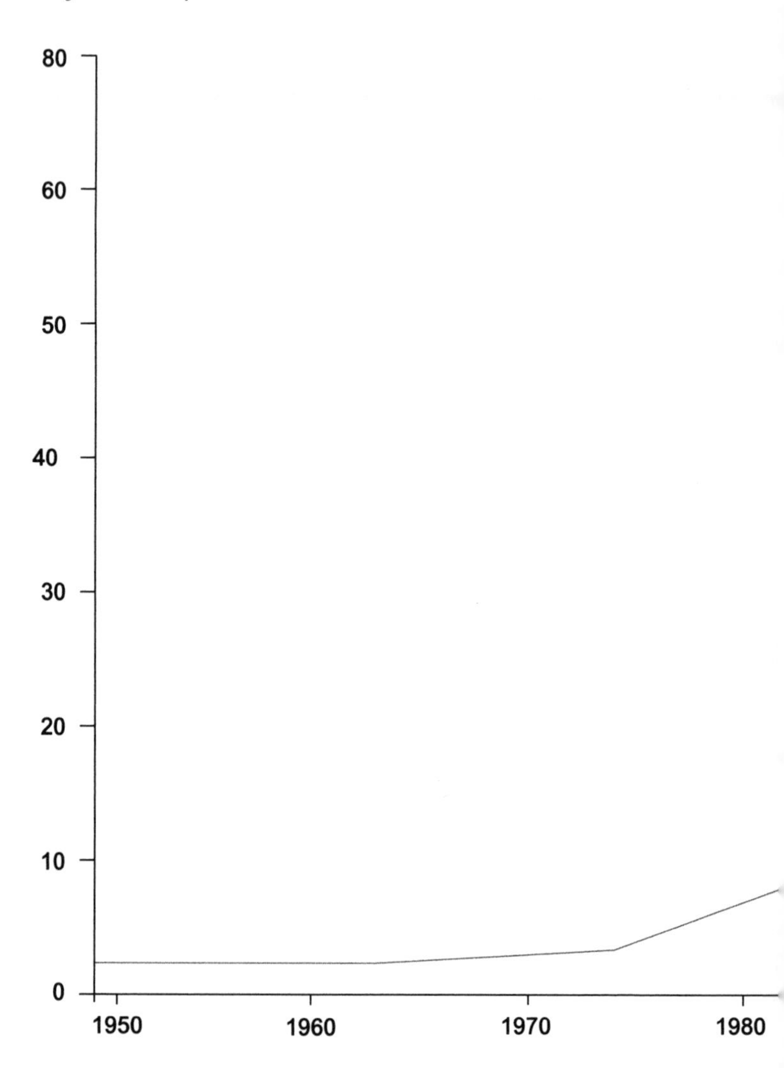

Major fires in the Americas are increasing rapidly. Between 1950 and 2000 the number of major fires on the American continent increased from two per decade to fifty per

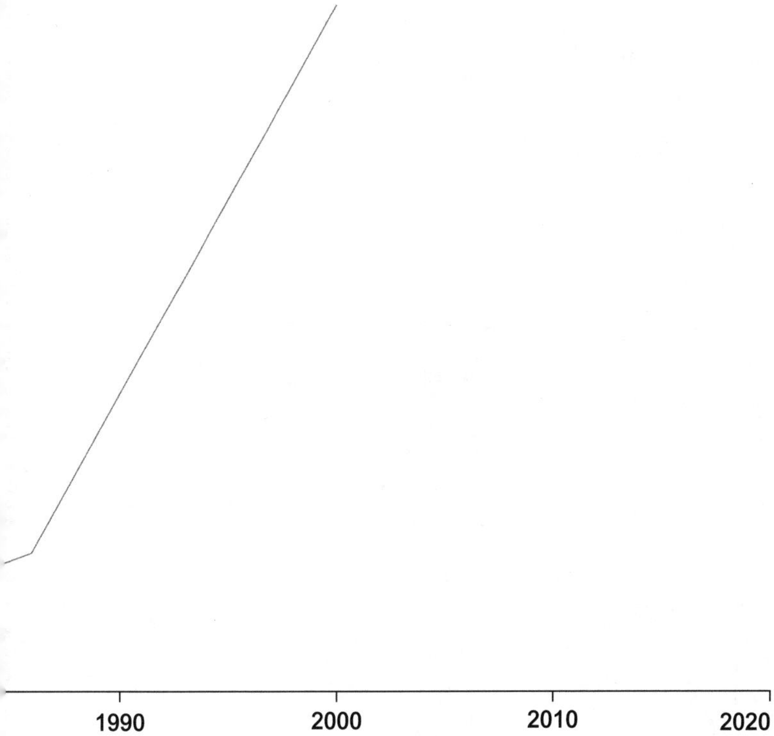

1990 2000 2010 2020

decade. Compiled from data from the U.S. National Interagency Fire Center, 2010; *Millennium Ecosystem Assessment Report,* 2005

Every which way you look at it,
a planet of ten billion looks like a nightmare.

And even more worryingly, there is now compelling evidence that entire global ecosystems are not only capable of suffering a catastrophic tipping point, but are already approaching such a transition.

What, then, are our options?

I can see two. The first is technologizing our way out of it. The second is radical behavior change.

Technologizing our way out of it

This is the domain of the "rational optimist." The rational optimist argument says that past predictions of doom—such as Malthus and Ehrlich—have turned out to be wrong, not least because our cleverness and our inventiveness have enabled us to technologize our way out of the population problem on every occasion. For example, we technologized our way out of a food crisis caused by rapidly rising population in the 1950s and 1960s by creating the green revolution.

Setting aside the fact that we've technologized our way into our problems in the first place, let's look at the current ideas for technologizing our way out of them. There are basically five ideas.

1. Green energy
2. Nuclear power
3. Desalination
4. Geoengineering
5. A second green revolution

Green-energy technologies refer to wind power, wave power, solar power, hydropower, and biofuels; they are often also called "renewables."

The fact is that current green-energy technologies are unlikely to be a viable planetary solution. It just isn't going to happen.

It is difficult to imagine existing green-energy technologies solving our energy demands at scale. For example, the next generation silicon photovoltaic cells require intensive mining of numerous metals and rare earths. Mining such metals involves processes that are anything but "green." Many of these metals are at severe risk of what is known in the trade as "cumulative supply deficits"—that is, they're running out. And the production of the new generation of solar panels involves nitrogen triflouride—one of the most potent greenhouse gases on Earth.

Secondly, even if green-energy technologies were a solution, which they're not, we would need to be embarking on a planetary-wide green energy program right now.

And we're not.

Even if we had embarked on such a comprehensive, worldwide program, which we haven't, it would be several decades before we could power the planet with green energy.

In the meantime, almost all of our energy will continue to come from fossil fuels—oil, coal, and gas—thus continuing to exacerbate the climate problem.

However, it is just conceivable that a radically new kind of green-energy revolution is possible.

Because plants have already done it. It's called photosynthesis.

If we could figure out how to do "artificial photosynthesis"—to learn from plants how to harness and convert the sun's energy—it might just be a potential global energy solution.

Only a handful of labs worldwide—such as mine—are even starting to think about this.

I never thought I'd say this, but nuclear power would seem to be the only existing technology that might solve the energy problem, at least in the short term—that is, for the next few decades.

But for nuclear power to be a solution, we would need to be embarking on a global nuclear power program right now.

And we're not.

Indeed, governments the world over are retreating from nuclear power, because it is expensive and politically unpopular, and because the commercial nuclear power industry does not want to pick up the longer-term costs of decommissioning and dealing with nuclear waste.

We could, potentially, solve some of the global water problem through the building of desalination plants, which convert sea water to usable water.

But again, we would need to be embarking on a massive desalination program right now. And such a program is not even on the horizon.

And even if it was—which it isn't—we would be solving one problem (the increasing shortage of fresh water) and exacerbating two others.

First, the energy problem; desalination is immensely energy-intensive. Second, the acceleration of coastal-ecosystem degradation, because desalination is extremely polluting.

Geoengineering is essentially the notion that planetary-scale engineering efforts might be needed simply to mitigate the worst consequences of the problems we face.

Some of the current ideas:

1. Seeding the global oceans with billions of tons of iron filings to accelerate the rate at which the oceans absorb CO_2. But this could be disastrous for marine life. And its effect on the global carbon cycle is completely unknown.
2. Building massive umbrellas in space to reflect the sun's energy back into space.
3. Putting aerosols into the atmosphere by either continually injecting sulfur dioxide into the atmosphere to mimic "daily" volcanic eruptions, or to inject limestone dust, titanium dioxide, or soot into the atmosphere to block solar radiation. In effect, to create "engineered" atmospheric

brown clouds (ABCs). I should point out that ABCs are already adversely affecting the health of some three billion people.

4. Carbon capture and storage (CCS), which is the concept of capturing carbon dioxide at source—power stations—and storing it underground. The world's largest "pilot" "CCS" project—in the UK—was canceled in 2012 because it proved technically and financially nonviable.

The problem is that all of the current geoengineering ideas are completely unproven.

All of them are extremely expensive. All of them come with significant knock-on effects, the long-term impacts of which are completely unpredictable.

I personally am not confident about geoengineering. In fact, I'm deeply skeptical.

As I mentioned earlier, there is currently no known way of feeding a population of ten billion.

So the idea of a second green revolution to solve this problem is now a really very hot issue.

We've had one green revolution, so the thinking goes: therefore, we should be able to create another.

What is certainly true is that we really do need a food revolution urgently. Because without one, billions of us are almost certainly going to starve.

But in exploring the idea of a second green revolution, it will pay us to look a bit more closely at the first.

The first green revolution focused on increasing crop yield. But to increase yield we had to introduce chemical fertilizers and breed shorter crops (that invest energy into increased production of seeds and flowers usable to us, rather than in growing taller).

In breeding shorter crops, we then had to compensate by deploying chemical herbicides to kill the weeds that would otherwise have grown taller than the crops and out-competed them for light.

We also bred out crops' natural defenses to pests, because plants' natural defenses to pests slow their rate of growth. But we then had to compensate for this in turn by introducing chemical pesticides.

We also bred crops to be ludicrously profligate with water, which we then had to compensate for by massively increasing water use for agriculture.

The green revolution was not a story about "clever people who worked out how to get more food from our fields." The truth is that the green revolution was a story about clever people thinking it was a good idea to buy every extra unit of food through energy and chemicals.

The green revolution was a myth.

We do need a food revolution. But it's a revolution that will require a radically new kind of science.

What else, in terms of technology?

What about new ideas and new technologies that might emerge at some unspecified time in the future? Certainly, the rational optimist's view is that our cleverness and inventiveness mean we don't have to worry: We will invent our way out of our current predicament. And—even I have to confess—it is immensely tempting to believe something so appealing. But it's a staggering leap into fantasy.

Given where we are, it would be very prudent, in my view, to be a rational pessimist right now.

So, as far as I am concerned, on today's evidence, technologizing our way out of this does not look likely.

So we need to do something else.

If we are not going to be able to technologize our way out of this, the only solution left to us is to change our behavior, radically and globally, on every level.

In short, we urgently need to consume less. A lot less. And we need to conserve more. A lot more.

To accomplish such a radical change in behavior would also need radical government action.

But as far as this kind of change is concerned, politicians are currently part of the problem, not part of the solution, because the decisions that need to be taken to implement significant behavior change inevitably make politicians very unpopular—as they are all too aware.

So what politicians have opted for instead is failed diplomacy. For example:

> The UN Framework Convention on Climate Change, whose job it has been for twenty years to ensure the stabilization of greenhouse gases in the earth's atmosphere: failed.
>
> The UN Convention to Combat Desertification, whose job it's been for twenty years to stop land degrading and becoming desert: failed.
>
> The Convention on Biological Diversity, whose job it's been for twenty years to reduce the rate of biodiversity loss: failed.

Those are only three examples of failed global initiatives. The list is a depressingly long one.

And, most recently, both the Rio+20 and the Doha COP (Conference of Parties) 18 produced even weaker rhetoric than all of these previous conventions, pledges, and "commitments."

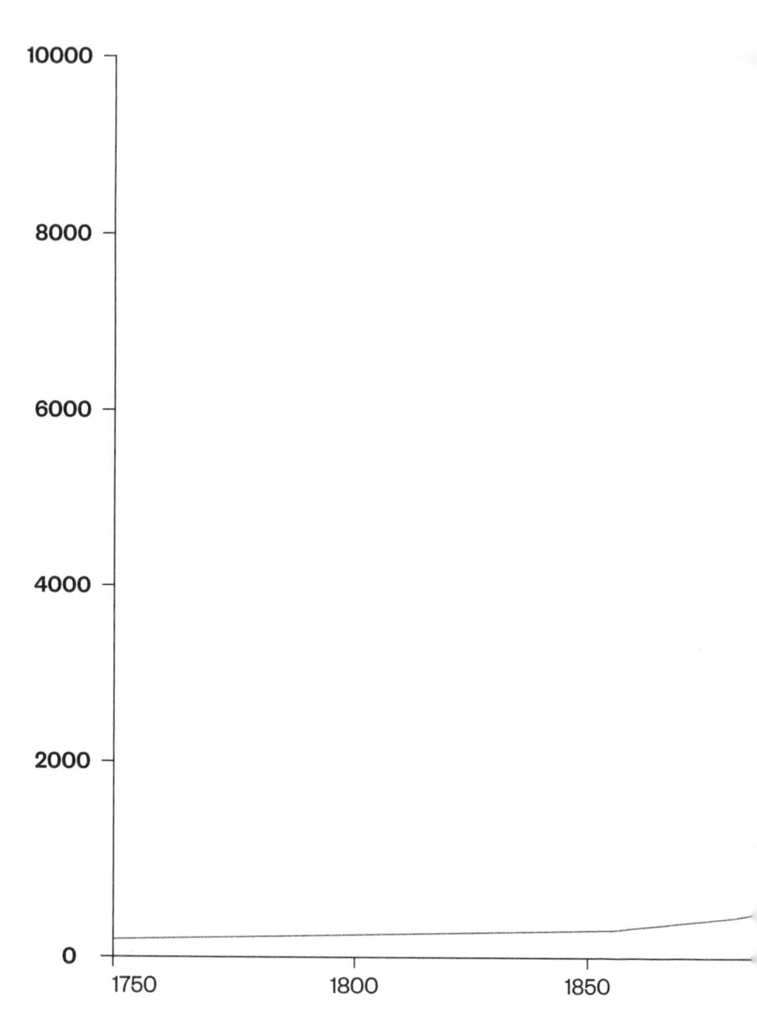

Despite twenty years of government pledges to tackle carbon emissions, we just keep on emitting more and more carbon. We are now emitting over 8.7Gt (Gigaton: billions of tons) of carbon per year through fossil fuel use. This does not even include CO_2 emissions from agriculture. Since the Kyoto Protocol (UNFCCC) was adopted in 1997, committing countries

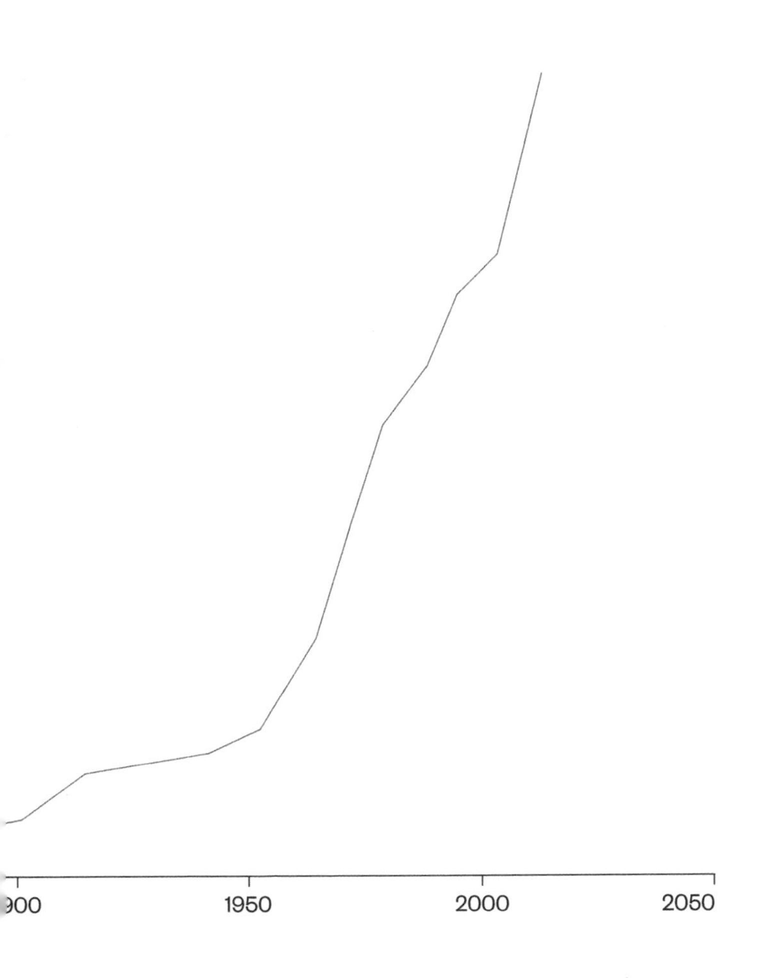

to reducing greenhouse gases, carbon emissions have risen from some 6.4Gt per year to 8.7Gt per year. That is a rise of 30 percent, in just fifteen years. And rising. Adapted from data from the U.S. Carbon Dioxide Information Analysis Center, Oak Ridge National Laboratory

It looks like twenty years of words and inaction are set to continue with another twenty years of words and inaction.

All the while, we are heading into deeper and deeper trouble.

And the way governments justify this level of inaction is by exploiting public opinion and scientific uncertainty.

It used to be a case of, "We need to wait for science to prove climate change is happening." This is now beyond doubt.

So now it's "We need to wait for scientists to be able to tell us what the impact will be and the costs." And "We need to wait for public opinion to get behind action."

But climate models will never be free from uncertainties.

And as for public opinion, politicians feel remarkably free to ignore it when it suits them. Wars, bankers' bonuses, and healthcare reforms, to give just three examples.

What politicians and governments say about their commitment to tackling climate change is completely different from what they are doing about it.

What about business? In 2008, a group of highly respected economists and scientists led by Pavan Sukhdev, then a senior Deutsche Bank economist, conducted an authoritative analysis of the value of biodiversity. Their conclusion? The cost of the business activities of the world's 3,000 largest corporations in loss or damage to nature and the environment, in terms of forest ecosystems that we lose each year alone, is estimated to be between $1.3 trillion and $3.1 trillion.

These costs are "externalities"—such as those I mentioned earlier about the cost of a car: the costs to society of business activities that are currently not being paid for. The cost of environmental damage, climate change, pollution, loss of ecosystems.

And an analysis just published estimates that the cost of a melting Arctic—which is now happening, and to which we are all contributing—will be as much as a staggering $60 trillion.

These are all costs that will have to be paid for in the future. By your children and your grandchildren.

G20 Leaders, London, 2009

In terms of business, the only hope of remotely mitigating some of the problems we are increasingly going to face is a radical transformation of corporate culture. To quote Pavan Suhkdev:

> "The rules of business urgently need to be changed, so corporations compete on the basis of innovation, resource conservation, and satisfaction of multiple stakeholder demands, rather than on the basis of who is most effective in influencing government regulation, avoiding taxes, and obtaining subsidies for harmful activities in order to maximize the return for just one stakeholder—shareholders."

Do I think that will happen? No.

What about us?

I confess I used to find it amusing, but I am now sick of reading in the weekend papers about some celebrity saying, "I gave up my 4 x 4 and now I've bought a Prius, aren't I doing my bit for the environment?"

They are not doing their bit for the environment. But it's not their fault. The fact is that they—we—are not being well informed.

And that's part of the problem. We're not getting the information we need. The scale and the nature of the problem is simply not being communicated to us. And when we are advised to do something, it barely makes a dent in the problem.

Here are some of the changes we've been asked to make recently, by celebrities (God help us), who like to pronounce on this sort of thing, and by governments, who should know better than to give out this kind of nonsense as "solutions":

> Switch off your mobile phone charger;
>
> wee in the shower (my favorite);
>
> buy an electric car (no, don't);
>
> use two sheets of toilet paper rather than three.

All of these are token gestures that miss the fundamental fact that the scale and nature of the problems we face are immense, unprecedented, and possibly unsolvable.

The behavioral changes that are required of us are so fundamental that no one wants to make them. What are they?

We need to consume less. A lot less. Less food, less energy, and less stuff. Fewer cars, electric cars, cotton T-shirts, laptops, mobile-phone upgrades. Far fewer.

Yet, every decade, global consumption continues to increase relentlessly.

And here it is worth pointing out that “we” refers to the people who live in the West and the North of the globe. There are currently almost three billion people in the world who urgently need to consume *more*: more water, more food, more energy. And by the end of the century there will be as many as five billion who will need to consume more.

Saying “Don’t have children” is utterly ridiculous. It contradicts every genetically coded piece of information we contain, and (at least in their conception) one of the most important (and fun) impulses we have. That said, the worst thing we can continue to do—globally—is have children at the current rate.

Even if a global nuclear power program were set up, even if geoengineering somehow took care of the climate-change problem, and even if we consumed less, we’d still at some point hit a brick wall if the human population continues to grow at anything like its current rate.

We all know there's a link between educating women in the developing world and reducing the birth rate. But despite this, and despite contraception being free in a number of countries where population is increasing, average birth rates are still three, five, or even seven children per woman.

According to the United Nations, Zambia's population is projected to increase by 941 percent by the end of the century. The population of Nigeria is projected to grow by 349 percent—to 730 million people.

Afghanistan by 242 percent,

The Democratic Republic of Congo
by 213 percent,

Gambia by 242 percent,

Guatemala by 369 percent,

Iraq by 344 percent,

Kenya by 284 percent,

Liberia by 300 percent,

Malawi by 741 percent,

Mali by 408 percent,

Niger by 766 percent,

Somalia by 663 percent,

Uganda by 396 percent,

Yemen by 299 percent.

Even the United States is projected to grow by 53 percent by 2100, from 315 million in 2012 to 478 million.

I do just want to point out that if the current global rate of reproduction continues, by the end of this century there will not be ten billion of us.

There will be twenty-eight billion of us.

Where does this leave us?

Let's look at it like this: If we discovered tomorrow that there was an asteroid on a collision course with Earth, and—because physics is a fairly simple science—we were able to calculate that it was going to hit Earth on June 3, 2072, and we knew that its impact was going to wipe out 70 percent of all life on Earth, governments worldwide would marshal the entire planet into unprecedented action.

Every scientist, engineer, university, and business would be enlisted: half to find a way of stopping it, the other half to find a way for our species to survive and rebuild if the first option were unsuccessful.

We are in almost precisely that situation now, except that there isn't a specific date and there isn't an asteroid.

The problem is us.

Why we are not doing more about the situation we're in—given the scale of the problem and the urgency—I simply cannot understand.

We're spending 8 billion euros (about 11 billion dollars) at CERN to discover evidence of a particle called the Higgs-Boson, which may or may not eventually explain the concept of mass and provide a partial thumbs-up for the "standard model" of particle physics.

And CERN's physicists are keen to tell us it is the biggest, most important experiment on Earth.

It isn't.

The biggest and most important experiment on Earth is the one we're all conducting, right now, on Earth itself.

Only an idiot would deny that there is a limit to how many people our Earth can support. The question is, is it seven billion (our current population), ten billion, or twenty-eight billion?

I think we've already gone past it. Well past it.

We could change the situation we are now in. Probably not by technologizing our way out of it, but by radically changing our behavior.

But there is no sign that this is happening, or about to happen.

I think it's going to be business as usual for us.

As a scientist, what do I think about our current situation?

Science is essentially organized skepticism. I spend my life trying to prove my work wrong or look for alternative explanations for my results.

I *hope* I'm wrong. But the science points to my not being wrong.

As I said at the beginning, we can rightly call the situation we're in an unprecedented emergency.

We urgently need to do—and I mean *actually* do—something radical to avert a global catastrophe. But I don't think we will.

I think we're fucked.

Image Credits

Grateful acknowledgement is given to the following for permission to reproduce images:

6–7	Apartment buildings in Shanghai. © Image Source/Corbis
26–27	Highway No. 1 Los Angeles, California, USA, 2003. © Edward Burtynsky, courtesy Nicholas Metivier Gallery, Toronto / Howard Greenberg & Bryce Wolkowitz, New York
50–51	Texas. From Eli McFadden's Terrain series. © Eli McFadden
56–57	Mirny, Yakutia, Russia, March 1996. © Jeremy Nicholl 1996
72–73	Soy plantation, Vilhena, Brazil. © Rodrigo Baleia/LatinContent/Getty Images
92–93	An atmospheric brown cloud shrouds Hong Kong. © Alex Hofford/epa/Corbis
96–97	VW Lot No. 1 Houston, Texas, USA, 2004. © Edward Burtynsky, courtesy Nicholas Metivier Gallery, Toronto / Howard Greenberg & Bryce Wolkowitz, New York
100–101	Crushed Cars No. 2, Tacoma 2004. © Chris Jordan
104–105	Burning tire pile, near Stockton, California, 1999. © Edward Burtynsky, courtesy Nicholas Metivier Gallery, Toronto / Howard Greenberg & Bryce Wolkowitz, New York
114–115	Twenty-four hours of air traffic. © Mario C. E. Freese—www.LX97.com
144–145	Food riots, Algeria, 2011. © STR/AFP/Getty Images
198–199	G20 leaders, London, 2009. © Eric Feferberg/AFP/Getty Images